Women IN STEM
Picking up *STEAM*

Researched by: Lauren Tate Baeza

Illustrated by: Arnelia Waters

© 2018 APEX Museum
All rights reserved

ISBN-13-978-1729670576

"My favorite exhibit was the Women In STEM exhibit. This served as a true inspiration for me. I was becoming a bit discouraged with attaining my Bachelor's in Math, as my graduation was recently pushed back. But standing in that room, I was surrounded with brilliant Black women and I felt comfortable- enough to cry in public as if no one else was there. And all I could keep telling myself was 'Britney, you can do this.' "

Student, Georgia State University

Table of Contents

Lilia Abron	2
Gloria Anderson	4
Wanda Austin	6
Alice Ball	8
Patricia Bath	10
Alexa Canady	12
Hattie Carwell	14
Jewel Cobb	16
Marie Daly	18
Etta Falconer	20
Njema Frazier	22
Evelyn Granville	24
Hadiyah Green	26
Euphemia Hayes	28
Fern Hunt	30
Shirley Jackson	32
Ashanti Johnson	34
Treena Livingston	36
Camara Phyllis	38
Norma Sklarek	40

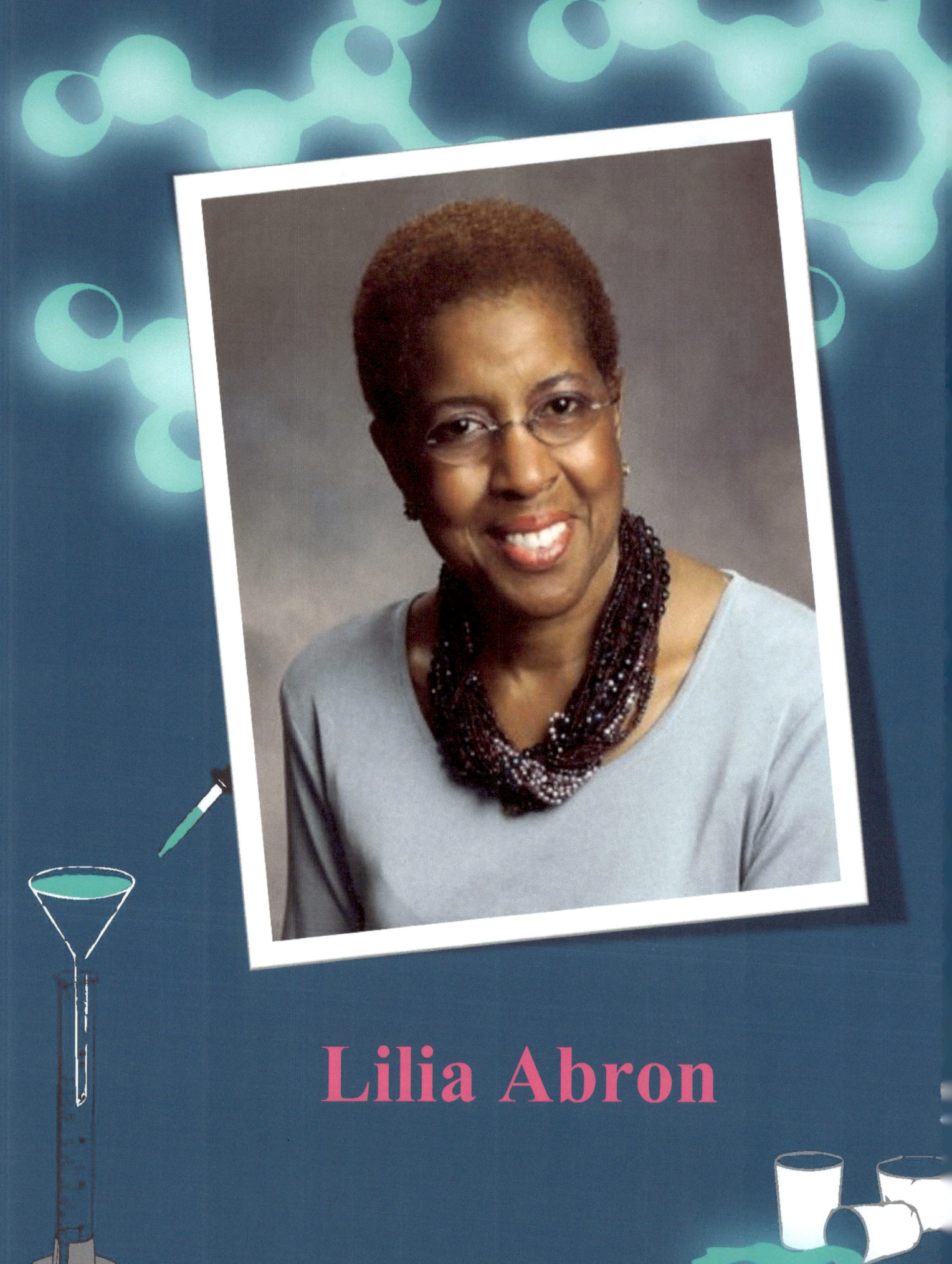

Lilia Abron

Lilia Abron

Lilia Abron is the first African-American woman to receive a Ph.D. in chemical engineering and the third woman to earn a doctorate in the discipline from the University of Iowa. She is Founder and CEO of PEER Consultants; a leading full-service environmental consulting firm in Washington, D.C. PEER Consultants has worked with federal, state, and municipal entities, as well as private sector clients for 40 years. Her research demonstrates the ways in which sustainability efforts can serve to strengthen communities and benefit the impoverished. Abron's team conducts surveys and inspections; analyzes potable water, wastewater, soil, and air samples; and manages water and wastewater treatment, collection, distribution, and disposal.

In 1994, Abron was one member of a small cohort of African-American engineers selected to help then newly elected president, Nelson Mandela, rebuild South Africa after apartheid. She founded PEER Africa Ltd and began building sustainable housing settlements and addressing many of the African nation's complex environmental concerns. With offices in 6 cities in the United States and 2 cities in South Africa, PEER Consultants works diligently to ensure clean and safe environments.

Gloria Anderson

Gloria Anderson

For several decades Gloria Long Anderson, Ph.D. has served as Fuller E. Callaway Professor of Chemistry at Morris Brown College, Atlanta, Georgia. Since coming to Morris Brown College in 1968, she has held many distinctive positions there including Fuller E. Callaway Professor of Chemistry, Dean of Science and Technology, Chair Academic Affairs Council, Vice President For Academic Affairs, Dean of Academic Affairs and from 1992-1993 she served as the college's Interim President. Additionally, she has contributed to the scientific community in many ways, holding posts as Research Consultant with The Hague in the Netherlands, Edwards Air Force Base, California, and as Research Fellow at Lockheed Martin Corporation, Georgia. She has instructed at Morris Brown College, South Carolina State College and Morehouse College, Atlanta, GA.

Dr. Anderson launched her career with a Bachelor of Science in Chemistry and Mathematics from Arkansas A.M. & N. College, Pine Bluff, Arkansas. She continued her studies in the field until earning a Ph.D. in Physical Organic Chemistry from The University of Chicago, where she did her dissertation on "19F Substituent Chemical Shifts of Bicyclic and Aromatic Molecules". She is a Certified Professional Chemist with extensive study and research on subjects such as "Potential Antiviral and Antitumor Agents", "Potential Replacements For Ozone Depleting Agents: The Synthesis of Some Hydrofluorocarbons," "New Synthetic Techniques For Advanced Propellant Ingredients", "Studies On The Mechanism of Epoxidation, 19F Substituent Chemical Shifts Of Bicyclic And Aromatic Molecules," "Ft-Ir Studies On The Curing Of Epoxy Resins among other listings.

In addition, Dr. Anderson holds substantial experience in the field of Public Broadcasting Policy and Management. She has served The Corporation For Public Broadcasting for many years and in many capacities, subsequent to her presidential appointment to the Board of Directors for the Corporation For Public Broadcasting (CPB), Washington, D.C. She has held numerous top positions in the broadcasting arena that have included state appointments, federal appointments and presidential appointments to serve. Capital among her many appointments is her service to the Food and Drug Administration, Rockville, Maryland. During her time there she made recommendations regarding orally inhaled and nasal drug products, ways to include inline process analytical technology methods to monitor quality control in drug manufacturing, regarding approval or disapproval of cardiovascular and renal drugs and dermatologic and ophthalmic drugs.

She is both an accomplished author and patent holder in her field of chemistry. Among her many publications are "Novel Synthesis of Some 3-Halo-1-Aminoadamantanes," "Educational Program Improvement in Chemistry Through the Acquisition of GC/MS and FT-NMR Instruments", etc. and her patented processes and methods are "1-Adamantyl Chalcones For The Treatment of Proliferative Disorders" and "A Method For Preparing Some 1-Adamantane-carboxamides,"etc.

Dr. Anderson is a member of Delta Sigma Theta and has a host of professional memberships including Golden Key National Honor Society Georgia, Sigma Xi Scientific Research Society, Beta Kappa Chi Scientific Honor Society, American Chemical Society, and many others. As well, she holds many honors, awards, and recognition on the county, state, regional and national level. Dr. Gloria Anderson is a nationally recognized African American female in STEM.

Wanda Austin

Wanda Austin

Wanda Austin is a system engineer recognized for her work on satellites and in defense. She served in several senior management and executive positions with The Aerospace Corporation, eventually becoming its CEO and president. The Aerospace Corporation produces technical and scientific research and advisory services to national space security programs. Austin serves on the Board of Directors for the Space Foundation and is a fellow of the American Institute of Aeronautics and Astronautics. She served on President Obama's Review of Human Spaceflight Plans Committee in 2009 and was appointed to the Defense Science Board and NASA Advisory Council.

Among numerous recognitions, Austin has received the National Intelligence Medallion for Meritorious Service, the Air Force Scroll of Achievement, and the Horatio Alger Award. During her 37-years with the Aerospace Corporation, the institute supported initiatives such as MathCounts and the United States' first robotics educational programs to encourage participation in math and sciences. The Dr. Wanda M. Austin STEM Scholarship, a four-year college scholarship that provides financial support and internships, is awarded annually to students pursuing these fields.

Alice Ball

Alice Ball

Alice Ball was a groundbreaking chemist and pharmaceutical researcher. Born in 1892 and raised by photographers, she was first introduced to chemistry through the picture development process. Ball was a top student at Seattle High School and began studies at the University of Washington in 1910 where she graduated with a bachelor's degree in chemistry in 1912 and a second bachelor's degree in pharmacy in 1914. While at the University of Washington School of Pharmacy, she published research, "Benzoylations in Ether Solution", in the esteemed Journal of the American Chemical Society. This publication granted her a full scholarship to the University of Hawaii, where she completed a Master of Science in chemistry in 1915. She was the first woman to graduate from the university and went on to become the first woman and first African-American instructor and researcher at the institution.

During this period in history, chaulmoogra oil was being used in China and India for the treatment of Hansen's Disease, commonly known as leprosy. However, the oil was not water soluble, which caused it to burn when injections were attempted and made it difficult to consume. Ball was the first person to develop a technique to extract the active ingredient in the oil. This came to be known as "The Ball Method." By producing an enhanced injectable treatment for leprosy, Ball achieved something that had eluded scientists for hundreds of years. In a 1918 article, the Journal of the American Medical Association reported that 78 leprosy patients had been released from quarantine, passing health examinations after receiving these improved chaulmoogra injections. Ball Method was used for decades to relieve the symptoms of leprosy.

Unfortunately, Ball passed away in 1916, at the age of 24, unable to witness the full scope of her accomplishments. Ball was demonstrating the use of gas masks when there was a malfunction. She became ill after inhaling chlorine gas and tragically did not recover. Ball was recognized posthumously by the Regent of the University of Hawaii in 2007, given its highest honor, the Medal of Distinction. In 2000, the Governor of Hawaii named a day in her honor. Fittingly, a bronze plaque rests at the base of the only chaulmoogra tree on the University of Hawaii's campus, a tribute to the school's first woman graduate.

Patricia Bath

Patricia Bath

Patricia Bath is an ophthalmologist, surgeon, inventor, and advocate for equity in healthcare. Her research on blindness led to the invention of the Laserphaco, a surgical tool that uses lasers to remove cataracts. Bath received a patent for the technology in 1988 and went on to hold three additional patents related to the Laserphaco. She was the first woman faculty member at UCLA's Jules Stein Eye Institute and the first woman to chair the Ophthalmology Department at Charles R. Drew University.

At the start of her career, Bath completed an internship at Harlem Hospital and a fellowship at Colombia University Eye Clinic. The disparities she observed between patients prompted her to conduct sociological research that demonstrated African-Americans were two times as likely to go blind and eight times as likely to develop glaucoma than white populations. Her study concluded that this was due to a lack of access to quality ophthalmological treatment. She co-founded the American Institute for the Prevention of Blindness, an organization committed to preventing blindness in less advantaged communities.

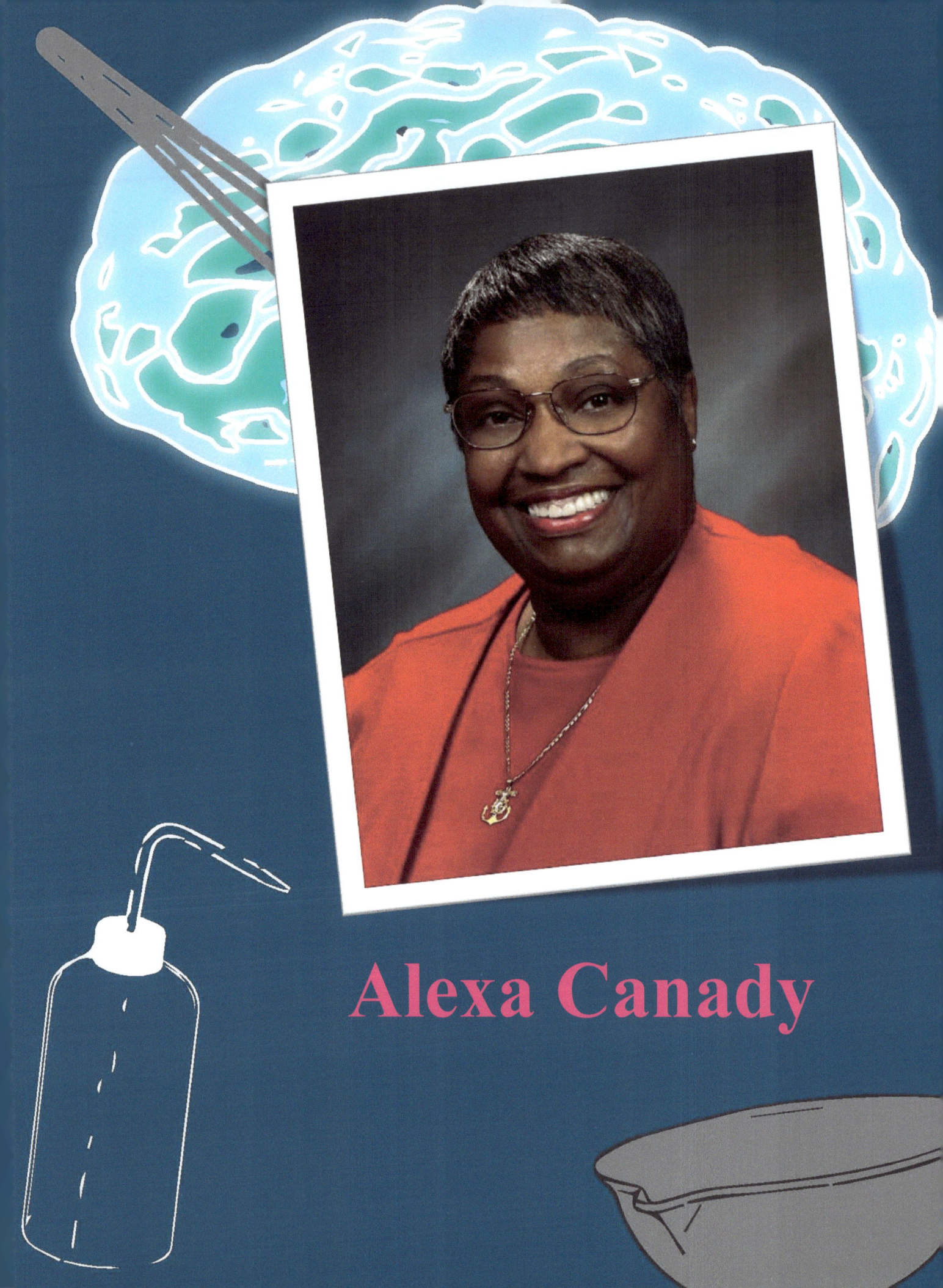

Alexa Canady

Alexa Canady

Alexa Canady became the first African-American woman neurosurgeon in the United States of America in 1984, being certified by the American Board of Neurological Surgery in 1984. Canady began her career as a surgical intern at Yale-New Haven Hospital and completed her residency at the University of Minnesota's Department of Neurosurgery. She was a fellow at the Children's Hospital of Philadelphia before joining the Children's Hospital of Michigan, where she would enjoy a long career and become the hospital's chief of neurosurgery. Canady performed surgeries to address head trauma, gunshot wounds, hydrocephalus, congenital spinal abnormalities, and other diseases impacting the brain.

As a pediatric neurosurgeon, she saved thousands of young children throughout the course of her career. Canady invented a programmable anti-siphon shunt to divert accumulated fluid in the brains of patients with hydrocephalus. She has been inducted into the Michigan Women's Hall of Fame, awarded the Distinguished Service Award from the Wayne State University Medical School. She holds numerous honorary degrees.

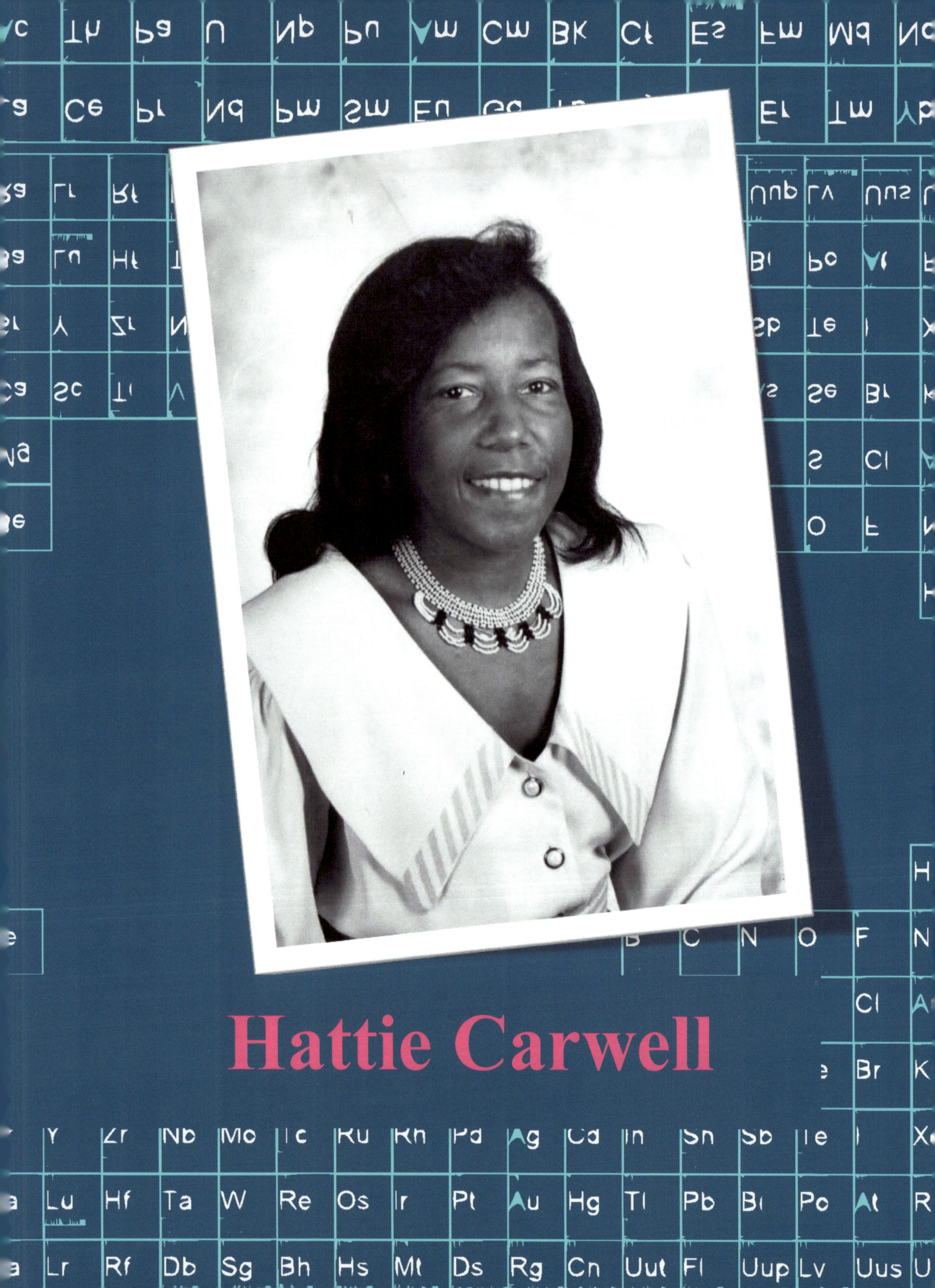

Hattie Carwell

Hattie Carwell

Hattie Carwell has served as President of the Northern California Council of Black Professional Engineers and Chair of the Development Fund for Black Students in Science and Technology. The fund has provided scholarships for hundreds of students pursuing a range of scientific disciplines. She is the author of Blacks in Science: Astrophysicist to Zoologist and Co-Founder and Executive Director of The Museum of African American Technology (MAAT) in Oakland California.

As a physicist with the United States Department of Energy for decades, Hattie Carwell worked to ensure that nations complied with international nuclear safeguard measures. Health physicist, Carwell inspected radiation safety at many of the Department of Energy's national laboratories. She became a senior engineer at the Department's offices in Berkeley, California, and later advanced from program manager of the High Energy and Nuclear Physics Program to operations lead and senior physical scientist at the Lawrence Berkeley National Laboratory, where she led initiatives in environmental health and safety oversight.

Jewel Cobb

Jewel Cobb

Jewel Plummer Cobb began her college career studying biology at the University of Michigan and Talladega College in Alabama. After completing her bachelor's degree at Talladega College, Cobb earned her master's degree and Ph.D. in cell physiology at NYU in 1947 and 1950, respectively.

Cobb became a cancer research fellow at Columbia University College of Physicians and Surgeons and the National Cancer Institute. She also served as Director of the Tissue Culture Laboratory at the University of Illinois from 1952 to 1954 and taught at various universities, including NYU, Hunter College, and Sarah Lawrence College. She later served as Dean of Douglass College at Rutgers University.

Cobb documented the role of hormones in cell division and published 36 articles on the cytology, growth, and genetic expression of both normal and cancerous cells. Her works are frequently cited by researchers and remain useful tools for scientists working to create more effective cancer-fighting treatment. She authored many papers and conducted significant research testing new chemotherapeutic drugs in cancer cells. In 1979 she published "Filters for Women In Science," which addressed discrimination and female representation in the sciences.

Cobb served as President of California State University's Fullerton campus from 1981 through 1990 making her one of the first African American women to lead a major four-year research institution in the Western United States. In this role, she acquired substantial funding for the Engineering and Computer Science Departments and is celebrated for diversifying the campus. After retirement, Cobb became a trustee professor at the California State University, Los Angeles.

She received the Lifetime Achievement Award from the National Academy of Science, the Kilby Lifetime Achievement Award, and the Reginald Wilson Award for her accomplishments in higher education, among other honors. Cobb greatly impacted the scientific community with her research, clearing a path towards equal access to professions for women and minorities.

Marie Daly

Marie Daly

Marie Maynard Daly graduated magna cum laude from Queens College in 1942, with a bachelor's degree in chemistry. One year later, she completed her Master of Science in the subject at NYU. She completed her doctorate in 1947 at Columbia University, where she investigated compounds and enzymes that aid in digestion. Her dissertation was entitled, "A Study of the Products Formed by the Action of Pancreatic Amylase on Corn Starch. Throughout her career, she would study the effects of cholesterol and sugars on artery health and heart function, as well as protein production in cells.

Upon completing her education at Columbia, Daly briefly taught physical sciences at Howard University. She then returned to New York to conduct research with the pioneering molecular biologist, Alfred E. Mirsky at Rockefeller University for several years. From 1955 through 1972, she would hold various key research positions for the American Heart Association, biochemist for the American Cancer Society, and cancer scientist at the Health Research Council of New York. Perhaps most notably, at the College of Physicians and Surgeons at Columbia University, Daly worked with the team that discovered the relationship between cholesterol and heart attacks.

She became a professor of medicine and biochemistry at Yeshiva University in 1960 and retired from the institution in 1986. In retirement, Daly served on the Commission of Science and Technology for the City of New York and established a scholarship for African-American science and medical students at Queens College. The scholarship was dedicated to the memory of her father, designed to prevent eager students from leaving school prematurely due to lack of funds.

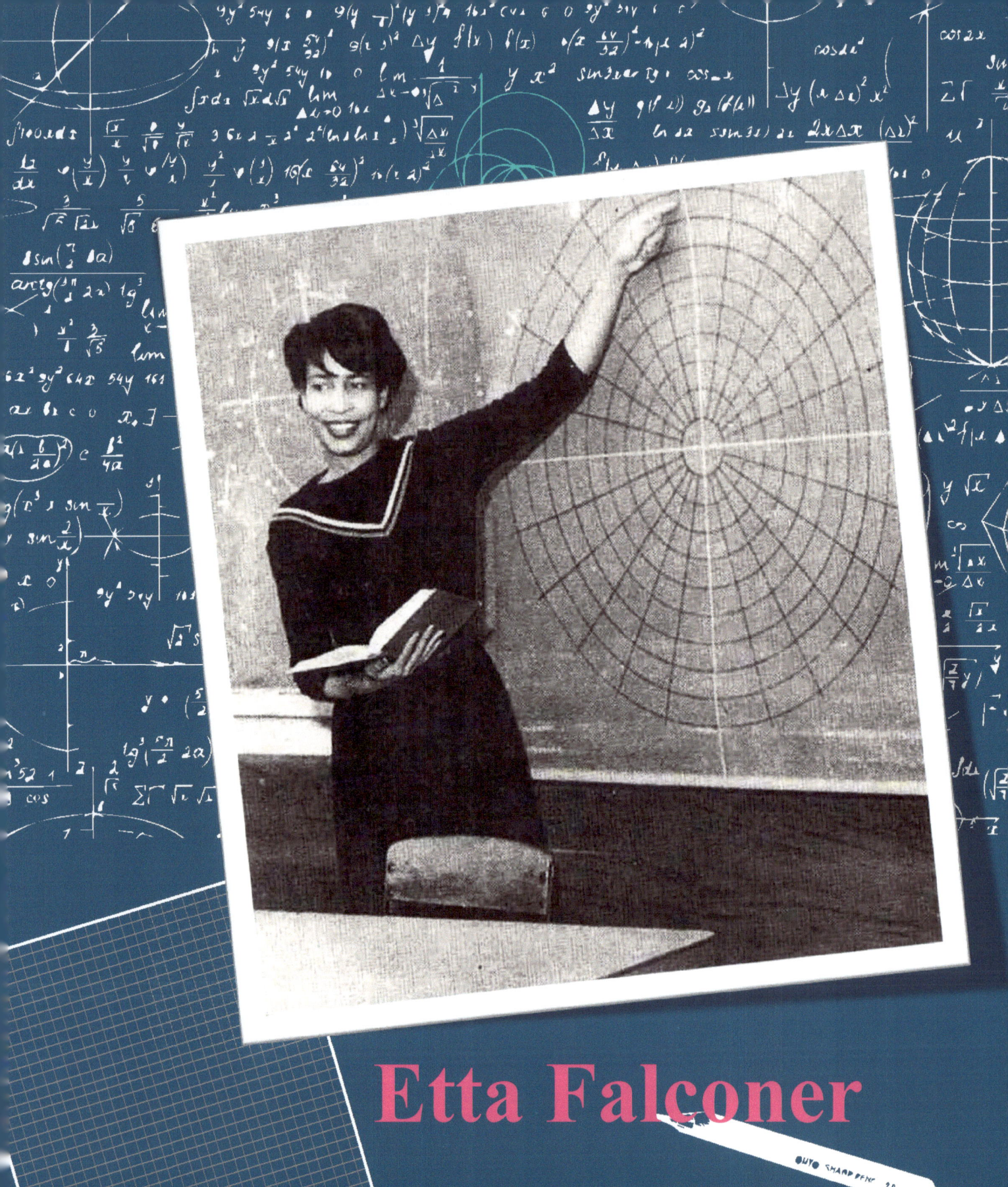

Etta Falconer

Etta Falconer

Etta Zubaer Falconer was born in Tupelo, Mississippi. Falconer was a leading mathematician and educator. She attended Fisk University in Tennessee, where one of her professors and early mentors was Evelyn Granville. Falconer completed her undergraduate degree in mathematics in 1953 and earned a Master of Science in the subject at University of Wisconsin, Madison in 1954. Upon graduation, Falconer taught at Okolona Junior College for nine years before joining the faculty at Spelman College, where she enjoyed a 30-year career. While teaching at Spelman, Falconer completed a Ph.D. in Mathematics at Emory University in 1969, becoming the eleventh African-American woman to do so. She completed her doctoral dissertation, "Quasigroup Identities Invariant Under Isotopy", which focused on algebraic loops and quasigroups. She demonstrated a lifelong commitment to learning and went on to earn a second master's degree in computer science from Atlanta University in 1982.

Falconer chaired both the Spelman Mathematics Department and the institution's natural sciences program. She later became Associate Provost of Science Programs and Policy in 1991 and held this post until retirement in 2002. At Spelman, Falconer instituted the NASA Women in Science Program and the NASA Undergraduate Science Research Program, geared towards preparing students for graduate school. Spelman College, a historically Black women's college, provided Falconer the opportunity to mentor and guide numerous women of African American descent towards careers in scientific fields. During her tenure, the university created chemistry and science departments, and STEM enrollment rose to represent 40 percent of all majors.

For her achievements, Falconer received a Lifetime Achievement Award from the American Association for the Advancement of Science. The organization specifically highlighted her commitment to helping young math and science scholars overcome social barriers presented by race and gender. She was the recipient of the National Science Foundation's Presidential Faculty Fellow Award, the Association for Women in Mathematics' Hay Award, and the United Negro College Fund Distinguished Faculty Award. For her invaluable efforts at the university, Falconer was honored with the Spelman Presidential Faculty Award for Distinguished Service.

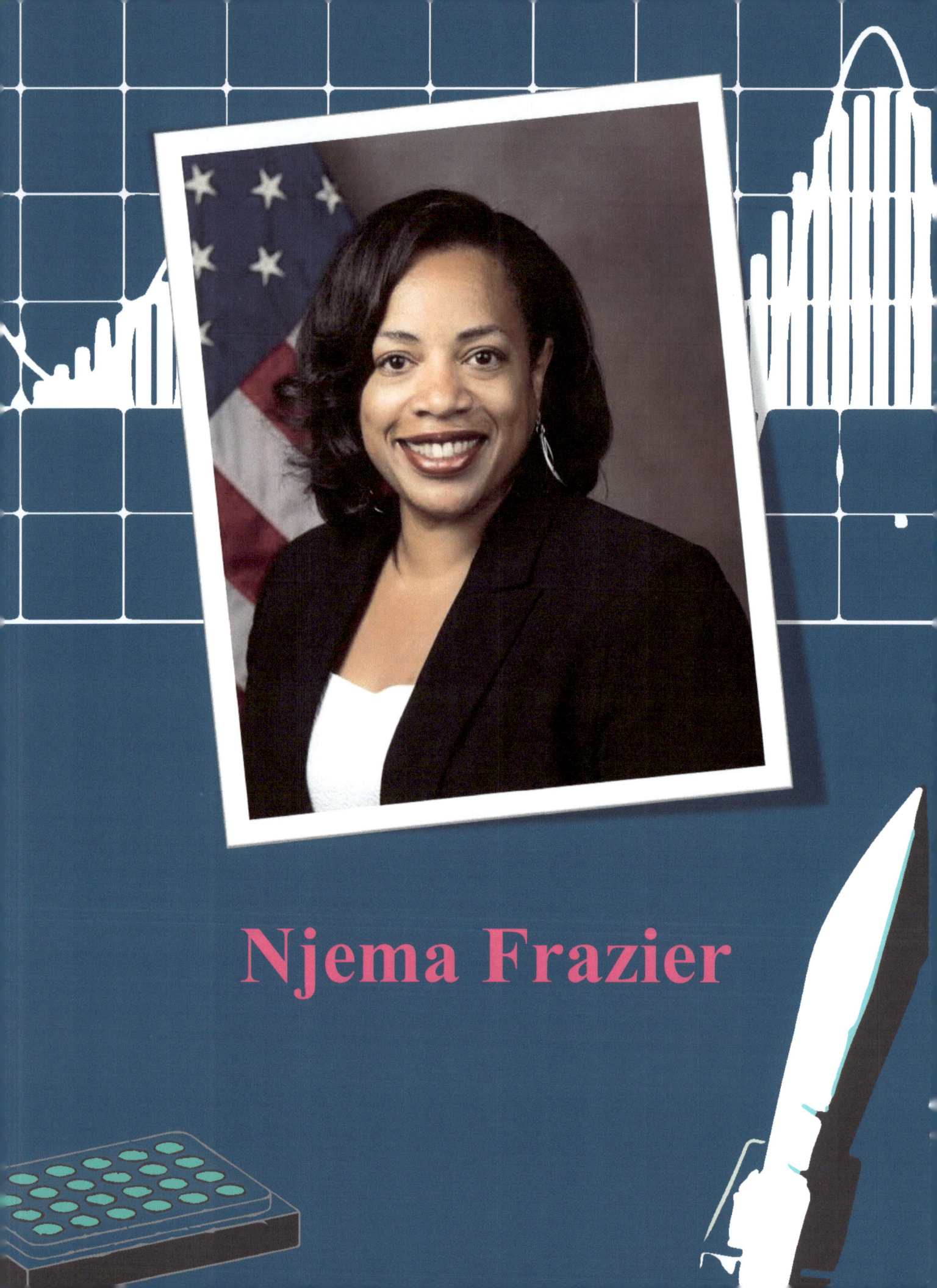

Njema Frazier

Njema Frazier

Njema Frazier is a theoretical nuclear physicist, leading technical efforts to ensure the United States remains a credible nuclear deterrent without nuclear explosive testing. Working with the United States Department of Energy in the National Nuclear Security Administration Office of Defense Programs, she oversees advanced nuclear weapons modeling and simulation. Frazier is Acting Director of the Inertial Confinement Fusion program, a program that examines the capabilities of high energy density physics. She is the first African-American woman to head the program in its 40-year history.

At the Department of Energy, she co-founded Professional Opportunities for Women at Energy Realized (POWER). POWER is an employee resource group designed to create a support system among women within the Department and to facilitate community outreach initiatives aimed at encouraging other women and girls to pursue careers in the sciences. Frazier has received the award for Distinguished Service to the National Nuclear Security Administration and was honored at the Black Girls Rock! Award show in 2017.

Evelyn Granville

Evelyn Granville

Evelyn Boyd Granville, a mathematician, made a substantial contribution to the United States defense and aerospace programs. With the aid of a scholarship from the African-American women's sorority, Phi Delta Kappa, she studied mathematics, theoretical physics, and astronomy at Smith College, graduating summa cum laude in 1945. Her 1949 graduation form Yale marked the second time an African-American woman completed a Ph.D. in Mathematics in United States history. Granville was a fellow at the Atomic Energy Commission and conducted postdoctoral research at NYU's Institute of Mathematics and Science. She briefly taught at Fisk University in Nashville, TN, where one of her students was **Etta Zuber Falconer**.

In 1952, Granville was hired by the National Bureau of Standards, a department later absorbed by the US Army and renamed Diamond Ordnance Fuze Laboratory. There she collaborated with scientists and engineers to work on missile fuses. In 1956 she accepted a position at IBM where she created computer software for NASA's Project Vanguard and Project Mercury space programs, working on calculations for satellite launchings and programming language, allowing her to develop programs for the 650 and 704 models.

Granville was part of the team of IBM scientists and mathematicians contracted by NASA to create and maintain programs to formulate trajectory and orbit computations. She conducted numerical analyses at the Federal Systems Division at IBM, the Computational and Data Reduction Center of US Space Technology Laboratories, and North American Aviation's Space and Information Systems Division, where she worked on calculations for the Apollo program.

Accepting an appointment as a professor of mathematics and computer programing at California State University Los Angeles in 1967, Granville returned to her foundation as a teacher until the 1980s. With her colleague, Jason Frand, she wrote Theory and Application of Mathematics for Teachers, a text used by some 50 universities. She retired as a math professor from the University of Texas in 1997. With a legacy of contributions to academia and aeronautics among her esteemed accomplishments, she was honored by the National Academy of Sciences in 1999.

Hadiyah-Nicole Green

Hadiyah-Nicole Green

Hadiyah-Nicole Green is a physicist specializing in nanobiotechnology. She is the first person to successfully cure cancer using nanoparticles. After witnessing her aunt and uncle struggle with and eventually succumb to cancer, Green embarked on a path dedicated to fighting the disease. Her research is centered on producing effective cancer treatments without the side effects of chemotherapy and radiation.

Trained as a physicist, Green crafted a treatment unspecific to the biology of cancer. She developed a laser technology that uses cancer-destroying nanoparticles to target unhealthy cells directly, in contrast to common cancer treatments which attack the healthy cells as well. In controlled tests and experiments, Green has cured cancer in mice using these therapies. Her goal is not to simply sell the treatment to pharmaceutical interests, but to ensure it is affordable to the average patient. Green, a professor at the Morehouse School of Medicine in Atlanta, GA, has been awarded substantial funding and support to further her research on this important patent-pending breakthrough.

Euphemia Hayes

Euphemia Hayes

Euphemia Lofton Haynes was the first African-American woman in the United States to earn a Ph.D. in Mathematics. She attended Miner Normal School in Washington, DC before completing her undergraduate education at Smith College. While at Smith College, Haynes studied mathematics and psychology, graduating in 1914. In 1930, she received a Master's in Education from the University of Chicago, where she also continued to study math. She later matriculated to the Catholic University of America and earned a Ph.D. in Mathematics. Her doctoral research concluded in 1943, with a dissertation titled "The Determination of Sets of Independent Conditions Characterizing Certain Special Cases of Symmetric Correspondences".

A lifetime resident of Washington, DC Hayes was dedicated to public education in the capital city. She became a math teacher at Armstrong High School and an English teacher at Miner Normal School, her alma mater. She then served as chair of the Math Department at Dunbar High School, an acclaimed and prestigious public secondary school for African-Americans in segregated Washington, DC. She later became professor and chair in the Division of Mathematics of Columbia Teachers College and founded the Math Department of the University of the District of Columbia, where she was a professor and head of the Department for 29 years. After retiring from the classroom, Haynes became the first woman to head the District of Columbia Board of Education, becoming its President in 1966.

Throughout her career, Haynes was an advocate for poor and minority students and a vocal proponent of equal access to quality education. She was a supporter of civil rights and an education activist, who fought against education discrimination. Specifically, Haynes was a critic of the district's racially biased track system, which disproportionately placed African-American and poor students in vocational programs as opposed to academic courses, despite their ability. The program was abolished under her term as President of the Board of Education. A devout Catholic, Haynes was awarded the Pro Ecclesia et Pontifice, a papal medal of honor, for her academic achievements and service. The distinguished educator was laid to rest in 1980.

Fern Hunt

Fern Hunt

Fern Hunt is a mathematician specializing in applied probability, stochastic geometry, and bioinformatics. Hunt's research investigated such phenomena as patterns in genetic mutation and mathematical expressions in nature. Working in both academia and for the federal government, the statistician has taught at the City College of New York, University of Utah, Howard University, and American University. She has also held positions with the National Institutes of Health and the National Bureau of Standards and Technology.

Hunt is recognized as instrumental in facilitating collaborative partnerships between mathematicians, scientists, and engineers to address problems that require layered, multi-disciplinary engagement. She serves on the Board of Trustees for the Biological and Environmental Research Advisory Committee for the Department of Energy. For her work using mathematics and computational biology to further understand genes, she was awarded the Arthur S. Flemming Award for Outstanding Federal Service.

Shirley Jackson

Shirley Jackson

Shirley Jackson is a theoretical physicist specializing in condensed matter physics, layered systems, optoelectronic materials, and quantum physics. She is the first African-American woman to earn a Ph.D. from MIT in any field and the second African-American to complete a doctorate in physics in the United States. In her early career, Jackson conducted research at Rutgers University and consulted on semiconductor theory at AT&T Bell Laboratories. She is considered a leading developer of Caller ID and Call Waiting on telephones. Jackson was appointed Chairman of the US Nuclear Regulatory Commission by President Clinton, a commission that safeguards nuclear reactor byproduct materials and ensures public health and safety. She served on the President's Intelligence Advisory Board.

President Obama awarded Jackson the National Medal of Science, the nation's highest honor for contributions to engineering and science, in 2016.

Ashanti Johnson

Ashanti Johnson

Ashanti Johnson is the first African-American to earn a doctorate in Oceanography from Texas A&M University. Using biogeochemical indicators, her research investigates and interprets the ways past events have impacted freshwater, estuarine, and marine habitats in the Arctic, as well as in coastal regions of the Caribbean and American South. She was previously Professor of Earth and Environmental Sciences at the University of Texas at Arlington and has taught at numerous other institutions, including the Georgia Institute of Technology, Savannah State University, and the University of South Florida.

Devoted to improving diversity in STEM careers, Johnson founded the National Science Foundation-funded Minorities Striving and Pursuing Higher Degrees of Success (MS PHD) Professional Development & Mentoring Institute and was formerly the Executive Director of the Institute for Broadening Participation. Johnson was awarded the Presidential Award of Excellence in Science Mathematics and Engineering Mentoring by President Barack Obama in 2009. She is currently CEO and Superintendent of Cirrus Academy Charter School in Macon, GA, a school specializing in preparing students to meet global standards in Science, Technology, Engineering, Art, and Math.

Treena Livingston

Treena Livingston

Treena Livingston Arinzeh is a bioengineer specializing in stem cell applications in orthopedic medicine. She previously worked as an engineer with both Merck & Company and Osiris Therapeutics Inc. Working with teams to develop cell-based regenerative therapies, Arinzeh developed FDA-approved methods to treat bone damage and defects. These methods heal more productively than bone graft therapies. Using the tissues of donors, she builds nano-scaffolds that can support new cells and encourage growth.

Her research has advanced to include the development and use of synthetic biomaterial to create scaffolds, an important alternative given the limited availability of donor tissue. Her work also investigates the repair of spinal injuries and damage. Arinzeh is a professor of Biomedical Engineering at the New Jersey Institute of Technology and has been the recipient of many prestigious honors, including the National Science Foundation Early Career Development Award and the Presidential Early Career Award for Scientists and Engineers, the nation's highest honor for young scientists and engineers.

Camara Phyllis

Camara Phyllis

Camara Phyllis Jones is a physician and social epidemiologist researching the impacts of racism on health. Her work examines unique environmental and social conditions endured by people of color, and the relationship between those conditions and health disparities and outcomes. She is acclaimed for bringing increased awareness to these determinants of health, consideration for which leads to more effective and comprehensive health policies and interventions.

An advocate for equal access to quality healthcare, Jones serves as President of the American Public Health Association. She previously taught at the Harvard School of Public Health and directed research on health and equity at the National Centers for Disease Control and Prevention. Jones has been the recipient of numerous honors and appointments, including a being a senior fellow in health disparities research at the National Medical Association Health Institute. She is currently a fellow at the Sather Health Leadership Institute, and a professor at both the Morehouse School of Medicine and Rollins School of Public Health at Emory University.

Norma Sklarek

Norma Sklarek

Norma Merrick Sklarek was a pioneering architect in Los Angeles and New York with a career spanning more than 30 years. Born in Harlem and raised in the Crown Heights neighborhood of Brooklyn by Trinidadian parents, Sklarek showed talent in the visual arts throughout her childhood and performed especially well in science classes at the New York public schools. She combined these interests upon entering the competitive architecture program at Columbia-affiliated Barnard College, culminating in a degree from Columbia University School of Architecture in 1950.

After receiving rejections from multiple firms, Sklarek took a civil service position at the New York City Department of Buildings. Disappointed but not defeated, she decided to take the state's rigorous architecture licensing exam, passing it on the first try, in 1954. With this, she became one of the first African-Americans licensed in the profession and the very first African-American woman to pass the New York State exam. Sklarek was hired by the acclaimed firm, Skidmore, Owings, and Merrill, in 1955 where she noted she found an environment of respect unlike the sexism and racism she had endured elsewhere.

Sklarek relocated to California in 1962 and, once again, became the state's first African-American woman licensed in architecture. In 1966, she was named a Fellow as a member of the Los Angeles chapter of the American Institute of Architects (AIA). She was the first woman to hold the honor and later became a Regional Vice President of AIA and chaired

Atlanta Public Art Audio Tour

404-260-5532

www.uforparents.org

A Mentor Can make a Difference

Photo Credit: Foundation For The Youths of Africa

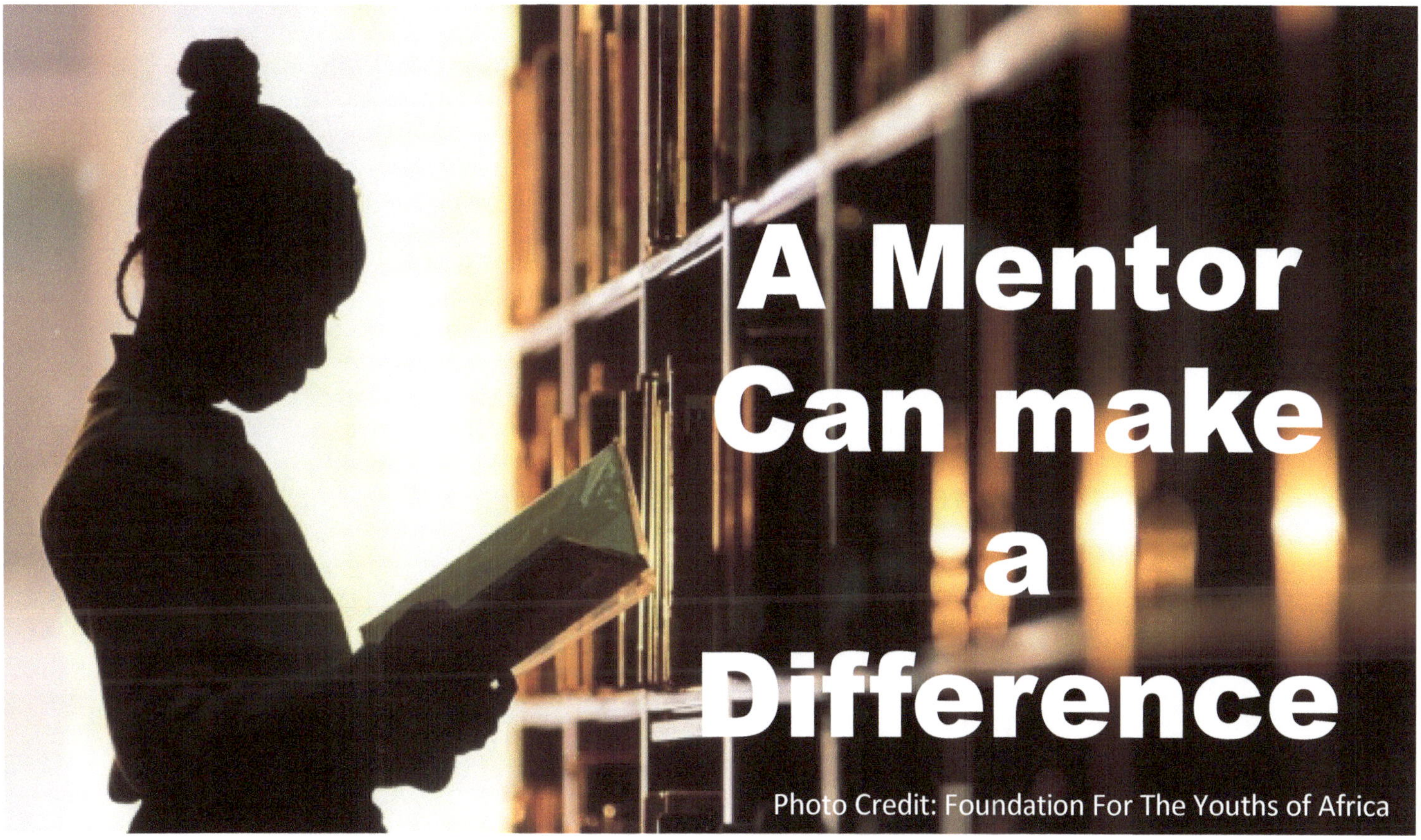

Atlanta CARES Mentoring Movement
An Affiliate Of The National CARES Mentoring Movement

www.Atlantacaresmentors.org

PARENTS SEND YOUR CHILD to College for Free

Successful Strategies that Earn Scholarships

TAMEKA WILLIAMSON

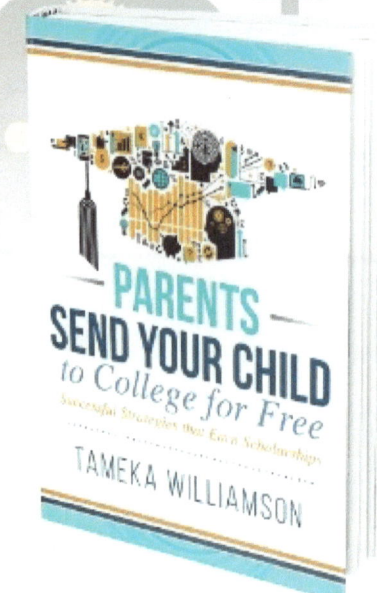

Tameka L. Williamson is an award-winning, multi-bestselling author, international trainer, speaker, coach and Executive Leadership Strategist who specializes in youth development and professional leadership. She is also part of the Forbes Coaches Council and John Maxwell Team. She has been featured in *The Huffington Post, Examiner, Lifetime Moms* and many other media outlets. Tameka currently lives in Atlanta, Georgia where she focuses on developing progressive leaders and high performing teams for today's marketplace.

ABOUT THE BOOK

Sending your child to college is possible, it starts with a plan, a strategy and a willingness to work the plan! Whether you are a college graduate or not, there is much to learn and do when it comes to getting your child college and career ready in the 21st century. *Send Your Child to College for Free* is the bestselling college planning guide that will help you position your college-bound child for college admissions and scholarships.

Providing practical strategies in an easy to read format, it offers simplified steps and advice to a complicated process, accompanied with action steps that produce results. Following Tameka's advice, you'll learn how to develop the college plan she used with clients that earned millions of dollars in scholarships. By following the advice in *Send Your Child to College for Free*, you'll discover the freedom and stress free life that comes with your child's education being covered with minimum to zero debt. As a result, families have garnered over $5 Million in grants and scholarships.

Free Scholarship eBook: CrushCollegeDebt.com

EdCauses

BUILDING STRONGER FAMILIES, SCHOOLS & COMMUNITIES: ONE PARTNERSHIP AT A TIME!

Grant Writing • STEM Teaching and Learning • Teacher Training • Family Counseling • Strategic Planning • Community and Business Development • Accreditation Planning • School Improvement

Dr. Lora Battle Bailey
Principal Partner

EdCauses.org
888-973-9337
partners@EdCauses.org

Ora Lee Smith
Cancer Research Foundation

Weareoralee.org

We can treat it! We can defeat it! "My goal is to change the way cancer is treated."
-Dr. Green

Help Us, Save Lives.

The Ora Lee Smith Cancer Research Foundation has the capacity to save lives and change the way cancer is treated. With our technology, we can. With your help, we will. Your support will help us translate a groundbreaking cancer treatment from the laboratory into the hospitals.

Be one of the 300,000 supporters we need to donate $100 or $10/month to reach our $30 million goal to cover the costs of clinical trials and FDA approval.

300,000 supporters x $100 = $30 million

Make your tax-deductible donation today of $100 or $10 a month, and "Become a Fundraiser."

Text **ORA** to **71777** to donate today!

Learn more at **www.WeAreOraLee.org**

Founder, Dr. Hadiyah-Nicole Green

Ora Lee Smith | Cancer Research Foundation

upscale

www.upscalemagazine.com

ENTERTAINMENT   BEAUTY & STYLE   LIFESTYLE   TRAVEL

Ultimate Success Experience

UPDATE
LATEST & GREATEST

Headline News

Entertainment

Lifestyle

Beauty

HIGHLIGHTS
HOT PICKS

Fashion

Travel

Cuisine

Business

MORE
SPOTLIGHTS

Contest

Giveaways

Videos

Shopping

MEDIA KIT   SUBSCRIPTIONS   CONTACT   NEWSLETTER   DIGITAL ISSUE

Don't miss a single moment   **LOG ON TODAY!**   @upscalemagazine

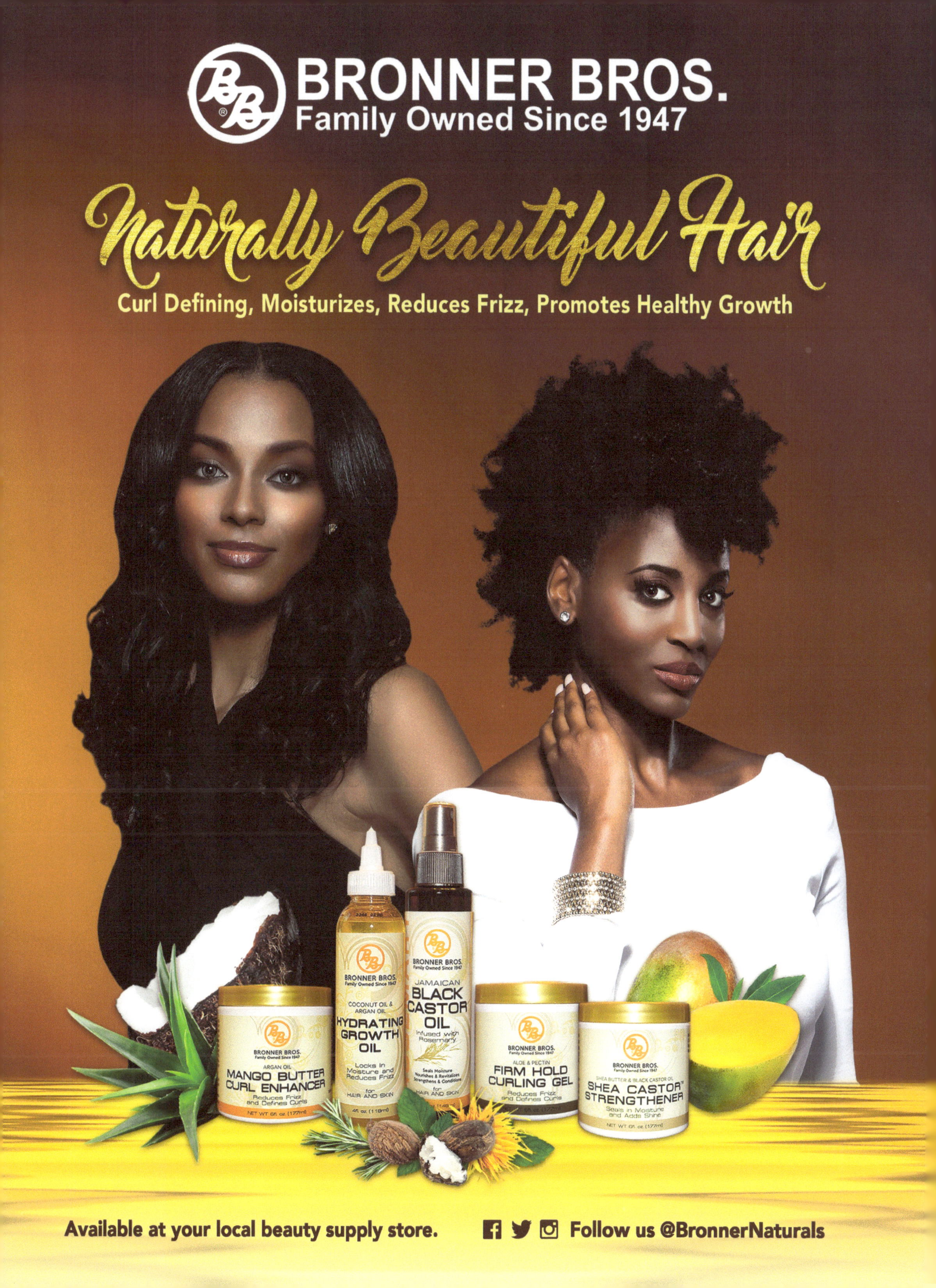
BRONNER BROS.
Family Owned Since 1947
Naturally Beautiful Hair
Curl Defining, Moisturizes, Reduces Frizz, Promotes Healthy Growth
BRONNER BROS.
ARGAN OIL
MANGO BUTTER CURL ENHANCER
Reduces Frizz and Defines Curls
BRONNER BROS.
COCONUT OIL & ARGAN OIL
HYDRATING GROWTH OIL
Locks In Moisture and Reduces Frizz
for HAIR AND SKIN
BRONNER BROS.
JAMAICAN
BLACK CASTOR OIL
Infused with Rosemary
for HAIR AND SKIN
BRONNER BROS.
ALOE & PECTIN
FIRM HOLD CURLING GEL
Reduces Frizz and Defines Curls
BRONNER BROS.
SHEA CASTOR STRENGTHENER
Seals in Moisture and Adds Shine
Available at your local beauty supply store.
Follow us @BronnerNaturals

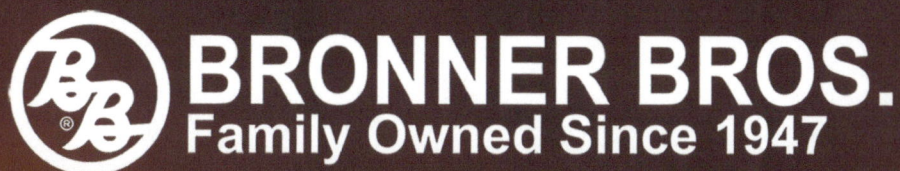

www.ingramcontent.com/pod-product-compliance
Lightning Source LLC
Chambersburg PA
CBHW051218220526
45473CB00003B/1084